I0491141

Ball Lightning caused the Chernobyl disaster

V. Torchigin

Copyright © 2019 V. Torchigin
All rights reserved.
ISBN: 9781653535262

Content

Introduction

The accident at the Chernobyl nuclear power plant is the most destructive of all known accidents in its consequences. Therefore, it is necessary to carefully study the causes of its occurrence so that such a tragedy does not happen again. Initially, the official version of the causes of the disaster came down to the mistakes of the operators who brought the station out of normal mode. Later, when Ukraine became an independent state, the official version was changed, and design flaws were identified as the main reason, since the design was carried out in Russia. Deficiencies in any complex system are always present. However, the question remains, were these deficiencies the real cause of the accident?

It is impossible to prove or disprove it. A simple example. There was a traffic accident on the highway, in which the car was in a ditch. Examining the causes of the accident, the traffic police found that the driver had significantly exceeded the speed, and his car was technically faulty: bad brakes, unacceptable play in the steering, etc. All this could lead to an accident. However, more thorough studies showed that immediately before the accident a truck was moving in the oncoming lane towards the car, and the driver had to turn into a ditch to avoid a collision.

The question arises. Was there any unknown reason similar to a truck moving in an oncoming lane? It is known from the chronology of events that approximately a 20 second before the explosion harmless preplanned electrical tests of the non-nuclear part of the station began. The tests began, and after 20 seconds the station was in ruins. The official versions do not establish any causal relationship between these events. There is also no such connection in the work of B.I. Gorbachev "Analysis of the causes and realistic scenario of the Chernobyl accident. The main choice between the two versions" [1].

Actual description of events before an explosion

It makes sense to provide some factual material and eyewitness accounts of these tests [1]. Electrical tests began at 01 h. 23 min. 04 sec. As the head of the pre-emergency evening shift of the fourth block, Y. Treguba, testifies,

"The run-out experiment begins. They disconnect the turbine from the steam and at that time they watch how long the run-out will last. And so the command was given .... We did not know how the run-out equipment works, therefore, in the first seconds, I perceived that "there was some kind of bad sound like that ... as if the Volga car started to slow down in full swing and started to move with skid. Such a sound: do-do-do-do, turning into a roar. There was a vibration of the building ... But not like an earthquake. If you count up to ten seconds - there was a rumble, the oscillation frequency fell. And their power grew. Then the blow sounded ... This blow was not very. Compared to what was then. Although a strong blow. The control room was shaking. I bounced, and at that time the second blow followed. Now it was a very strong blow. Stucco fell, the whole building shacked ... the lights went out, then emergency power restored ... Everyone was in shock ... "[1].

Here is how B.V. Rogozhkin describes the course of the accident, who worked on an emergency night as the shift supervisor of the station. "All events took place within 10-15 seconds. Some kind of vibration appeared. The hum rumbled rapidly. The power of the reactor first fell, and then began to increase, not giving in to regulation. Then - a few sharp pops and two "hidro shocks". The second more powerful - with side of the central hall of the reactor. On the block billboard the lights went out, the slabs of the suspended ceiling fell, all the equipment turned off" [1].

The shift supervisor Orlov recalls that at first he heard a rumble similar to the sound emanating when one of the parts of the turbogenerator went out of operation, Then, according to him, a fire broke out at the lower elevations - transformer oil was burning. Other employees of the power plant recall the same thing.

Thus, from the description of the development of events, we can draw the following conclusion

1. the start of testing, initiated appearance the following phenomena observed for 10-15 seconds

2. there was a bad sound like do-do-do-do ....

3. the repetition of this sound decreased, but the strength of the sound increased. After that there was an explosion.

4. the smell of burned transformer oil was felt

The employee of the Kurchatov Institute, Doctor of Physical and Mathematical Sciences L. I. Urutskoev in the article "Chernobyl may happen again. The mystery of a nuclear disaster is close to a solution" [2]. noted the following two very strange picture when examining the reactor mine.

5. "a great circle of **a regular round shape**" in a sheet of the metal wall of the reactor was burnt by unknown object

6. the walls of the reactor shaft of the 4th power unit were covered with an oil emulsion".

The official version of the cause of the accident

In the official version, these circumstances are considered a trifle and are not analyzed. There are many reasons that could lead to an accident, which are described in detail on Wikipedia. In particular, it is indicated that the cause of the accident could be the destruction of heat-generating elements or fuel channels.

For a better understanding of the causes of the accident, Figure 1 shows a scheme of the atomic station. In accordance with this scheme, there is a sealed system where water is heated when passing through heat-generating elements 3 (relatively cold blue water when passing through heat-generating elements is heated to a temperature of about 300 ° C and is shown in red in the figure).

A mixture of water and steam is in a sealed system under a pressure of about 70 atmospheres. In devices 6 steam is separated from the water and enters the turbines that rotates the generators of electrical energy. After the turbines, the steam is condensed in a cooler 13 that uses water from a nearby reservoir. This relatively cold water enters the bottom of the pressure channels with fuel rods 3 again.

According to the official version, the cause of the accident is formulated as follows: "It is not known for certain why the power surge began, which led to the destruction of the Chernobyl reactor. Only a general idea of the scenario of the accident. Unified in authoritative versions is an uncontrolled increase in reactor power. The destructive phase of the accident began when fuel elements in a certain area in the lower part of the reactor core were destroyed due to overheating of nuclear fuel. This led to the destruction of the shells of several channels 3 in which these fuel elements are located, and steam under a pressure of about 70 atmospheres got access to the reactor space, in which atmospheric pressure of 1 atmosphere is normally maintained. The pressure in the reactor space increased sharply, which caused further destruction of the reactor as a whole.

At the same time, the authors of the official version cannot even answer the question of whether the initial overheating and destruction of fuel rods occurred due to a sharp increase in the reactor power due to the appearance of a large positive reactivity in it, or vice versa, the appearance of positive reactivity is a consequence of the destruction of fuel rods that occurred by any another reason?

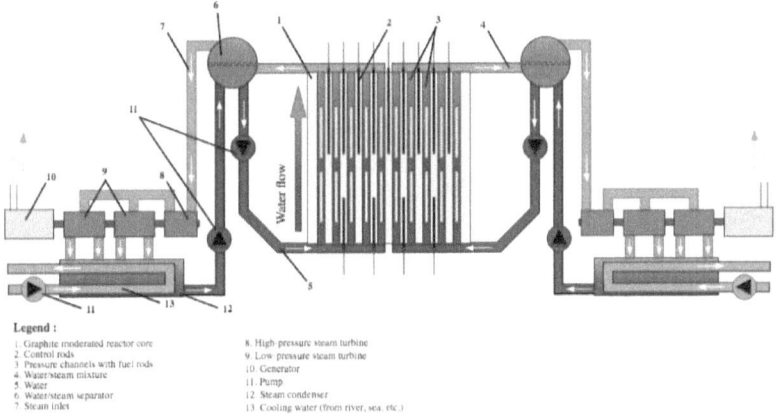

Fig.1. Scheme of the Chernobyl plant

Of course, without analyzing all existing objective data, hundreds of different hypotheses and versions can be put forward. Some of them are listed on Wikipedia. In order to restore the true cause of the accident, the above 6 reliably established irrefutable circumstances should be explained on the basis of the known laws of physics.

Preliminary analysis of the above presented data

It is safe to say that immediately after the start of the test, relaxation oscillations appeared in the system. Such oscillations are observed in nature very often. For example, each leaf of a tree, being in the wind, which has constant speed and direction, is not stationary, but oscillates, making relaxation oscillations. The same can be said about the branches and tops of trees. The creaking of doors, the squealing of brakes, the sounds made by wind musical instruments are also caused by relaxation oscillations. Relaxation oscillations are also responsible for sounds when a person speaks or sings. At the same time, if desired, a person can release air from the lungs without sounds, that is, he can control relaxation oscillations.

Closest to the sounds of do-do-do-it is the sound of a working pneumatic jackhammer. In this case, there is a sealed system where air is compressed under pressure of several atmospheres. When the hammer is pressed on an obstacle, the impermeability is broken and the hammer makes sounds do-do-do-do.

Probably, everyone had the opportunity to observe the consequences of such relaxation oscillations occur in a sealed system in the form of ordinary water supply in an apartment building. Sometimes, when the valve opens, a sound is heard, reminiscent of a buzzle, crash, as well as do-do-do that residents of other apartments hear. The main reason for the appearance of such sounds is a violation of the normal mode of operation of one of the many components of the water supply system that is also airtight and under pressure of 2-3 atmospheres.

Relaxation oscillations are also responsible for sounds when a person speaks or sings. At the same time, if desired, a person can release air from the lungs without sounds, that is, he can control relaxation oscillations. Relaxation oscillations are more common in nature than their absence. We must try to get rid of relaxation vibrations.

The nature of relaxation vibrations can be most clearly illustrated by the following example. Let the rod rest on a table and can be driven by an elastic spring, as shown in Fig. 2. It is known that the static friction between the bar and the table is greater than the sliding friction. When the spring is gradually stretched, the bar remains at rest until the spring tension becomes greater than the friction force at rest.

When the tension of the spring exceeds the friction force at rest, the bar begins to move with acceleration, since it is affected by a force equal to a

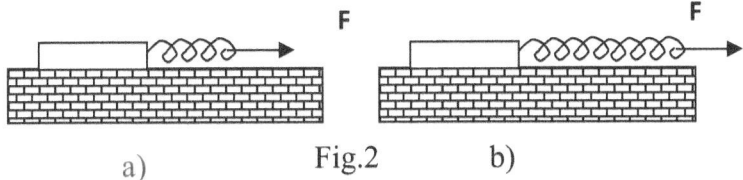

difference between the rest and sliding friction forces. In this case, the elastic tension gradually decreases, the tension force disappears and the speed of the bar gradually drops to zero. The bar stops. The process is then repeated. As a result, at a constant speed of the right end of the spring the bar moves irregularly.

If the spring elastic is weaker (for example, its length is doubled as is shown in Fig.2 b), the spring is stretched to the doubled length before the bar begins to move. As a result, the frequency of jumps decreases by two times, and the distance that the bar moves with each jump increases by two times.

A similar situation occurs when a car brakes sharply, so that the wheels stop spinning and the car slides along the road. We hear the screech of the brakes. This sound is created by relaxation oscillations of tires when sliding on asphalt. A whistle corresponds to a high frequency of several hundred hertz.

On the contrary, if the car moves in a direction perpendicular to its axis, then a low-frequency sound such as do-do-do-do is heard. This is due to the fact that the braked wheels of the car are fixed in the transverse direction less rigidly than in the longitudinal direction (since the transverse movement of the car is exotic). The amplitude of the deviation of the car from the usual position relatively the road at motionless wheels in the transverse direction is much greater than the amplitude of the deviation of the car in the longitudinal direction.

One can imagine a situation when a car with brake wheels is towed sideways so that its axis is perpendicular to its movement. In this case, the sound "do-do-do-do" will be heard. The wheel mount in the transverse direction will weaken, as it is not designed for such a movement. In this case, the frequency of do-do-do-do will gradually decrease, and the amplitude of the jumps will gradually increase. In the end, an accident will happen. The wheels will separate from the car.

Perhaps the situation is more familiar when a conventional table is moved across the floor by means of pushing. This movement is often accompanied by sounds such as "do-do-do-do". As a result of this movement, the attachment of the table legs to the countertop is gradually weakening. The weaker the table legs are attached to its tabletop, the greater the amplitude of the relaxation oscillations, and the smaller their frequency. Finally, the legs of the table are separated from the table, that is, the destruction of the table occurs.

A similar picture was described by Orlov after the start of electrical tests before the explosion. At first, "some kind of bad sound appeared, as if the Volga car started to slow down in full swing and started to move with skid. Such a sound: do-do-do-do ... "The oscillation frequency was falling, and their power was growing". These oscillations led to the failure of the system. "The blow sounded."

Thus, a primitive analysis at the high school level leads to the conclusion that the main cause of the accident was a violation of the sealed system, where water and steam are at a pressure of about 70 atmospheres. This violation gradually increased, which ultimately led to the depressurization of this system. Further developments no longer apply to the original cause.

Taking this irrefutable circumstance into account would have avoided most of the interpretations associated with operator errors and design flaws. Such deficiencies can be investigated regardless of the specific circumstances of the accident.

Bearing in mind the above considerations, the problem of finding the root cause of an accident is much simpler. It is necessary to find the reason for the occurrence of relaxation oscillations or the reason why the sealed system under 70 atmospheres pressure exited its normal mode of operation.

How could the harmless fact of disconnecting the turbine from the steam affect the nuclear reactor, bearing in mind that the turbine is not in close proximity to the reactor? Studying recently the properties of ball lightning and the reasons for their appearance, we came to the conclusion that there is good reason to believe that the cause of the accident was ball lightning, which occurred when the generator was disconnected from steam during high current switching.

In this publication, we show that all above mentioned circumstances could have occurred at an arise of an ball lightning at the moment when the turbine was disconnected from the steam, which was accompanied by switching great currents.

To substantiate this hypothesis, it is necessary to show the possibility of existence of the following events

1. the appearance of ball lightning when switching high currents;

2. suction into the ball lightning of vapors of transformer oil, the combustion of which was noted by witnesses.

3. penetration of ball lightning into the steam line (into a closed system under pressure of 70 atmospheres)

4. moving ball lightning into the mine

5. burn holes in the wall of the mine

6. the disappearance of ball lightning with the release of vapors of transformer oil from it and the condensation of these vapors on the walls of the mine.

Initially, based on the known properties of natural ball lightning given in literary sources, we present known properties of ball lightnings that can be responsible for arising events

To give our arguments greater credibility, we give APPENDIX in which we show a possibility of arising the same events on the basis of the optical theory of ball lightning. To show that this theory correctly describes the physical nature of ball lightning, an explanation of almost all known anomalies of natural ball lightning is presented. Based on this theory, we show that all above mentioned conditions can be caused by ball lightning. The theory has been known since 2003. Since that time, more than 30 articles have been published in leading international journals in physics and optics, which explain on the basis of this theory all the anomalous properties and the mysterious behavior of natural ball lightnings.

Specific properties of ball lightnings

Justification for the possibility of arising ball lightning does not cause difficulties. It is well known that ball lightnings occur most often in battery submarines when reversing engines. Such ball lightning flies through cabins and sometimes burn racks and burn partitions. Numerous evidence of the appearance of ball lightnings is described in Barry's monograph [3], devoted to ball lightning. For example, Norwegian hydro engineer Nielson showed that a reddish luminous mass separated from the circuit area of the DC generator and remained visible in the air for several seconds. Photographs of the luminous ball separated from the luminous mass are given.

A significant contribution to the study of the properties of ball lightning and the conditions favorable for their production in a laboratory was made by R.F. Avramenko and his colleagues. In 1994, they published the book "Ball Lightning in the Laboratory", which presents the results of studies of so-called autonomous objects (AO) obtained by an erosive gas discharge [5].

The erosive gas discharge differs from the usual one in that during the discharge, the substance (usually hydrocarbons) evaporates and is absorbed into the shell of the luminous object. The positive effect of gas erosion discharge is confirmed in numerous experiments. The beneficial effect of erosive gas discharge on obtaining luminous anomalous objects was found experimentally. Explanation of the beneficial effect of the erosive gas discharge on the appearance of luminous anomalous objects was given as part of the optical model of ball lightning.

When trying to obtain artificial ball lightning in a laboratory, it was experimentally shown that the so-called "erosive gas discharge" is favorable for obtaining luminous objects with anomalous behavior that resembles the behavior of natural ball lightning.

In an erosive discharge, the electrode or substance deposited on the electrode evaporates. This method of producing AO was subsequently used by many researchers [6, 7]. The favorable effect of an erosion discharge on the occurrence of AOs was explained recently, after the physical nature of ball lightning was explained and it was shown that AOs are miniature ball lightnings with a relatively short lifetime [8]. Any substance can be used as a substance that undergoes erosion during a gas discharge. For example, in experiments on the production of artificial ball lightnings, wax was used as such a substance, which was previously deposited on electrodes between which a gas discharge occurred [12]. In other experiments, metal, various polymers, cotton wool, wood shavings were used as such a substance [13].

Erosion gas discharge occurs in any oil circuit breakers. As a result, the oil gradually loses its properties and is periodically replaced. Of course, measures were taken in the design of the switch so that ball lightning did not occur. In most cases, these measures work. However, as evidence of the occurrence of ball lightning in submarines, they show that there are exceptions. Perhaps such an exception occurred at the beginning of the test.

In an erosive gas discharge, erosion products are retained in ball lightning. For example, Stakhanov [4] describes a case where linear lightning struck a tree and ball lightning popped out of it. After its disappearance, a yellow hot mass remained. The observer touched it and burned his fingers.

It can be assumed that ball lightning occurred during an erosive gas discharge in which transformer oil was subject to erosion. Vapors of this oil were compressed in a ball lightning shell. With the disappearance of ball lightning in the reactor shaft, vapors of transformer oil were condensed on the walls of the shaft. As noted by Urutskoev, the walls of the mine were covered with an oil film.

Thus we can assume that switching great currents is accompanied by an erosive gas discharge. As a result of this discharge, the transformer oil evaporated, the vapors of which were captured by ball lightning.

The possibility of penetration of ball lightning into the steam line is confirmed by numerous reports of the possibility of penetration of ball lightning into confined spaces. For example, in the book of Sagan, an entire chapter is devoted to the description of cases of penetration of ball lightning into the airliners

Besides, the following evidence is given in Stakhanov's book [4].

During a severe thunderstorm, ball lightning of 20-30 cm in diameter entered through a hole in the wall for grounding.

During thunderstorms, ball lightning 10 cm in diameter penetrated in the hole 2 cm wide. Ball lightning deformed in "stretched sausage." while penetration.

Ball lightning 10-20 cm diameter went into the crack around the window glass.

Ball lightning diameter of 30-50 cm entered through a small hole in the window (glass chipped corner) 1-1.5 cm in width as the "yellow thread".Having done a few laps around the room, ball lightning exploded after 20-30 seconds.

Ball lightning 5-10 cm in diameter entered as a "snake" during a thunderstorm through the open window and then forming a bead. After going around the room a distance of 5-10 m, ball lightning disappeared without an explosion near the switch.

During a severe thunderstorm, ball lightning entered the house in the gap between the boards around the pipe. The board was smoked. The fire began.

Ball lightning 10-20 cm in diameter passed through a hole diameter of 8 cm.

Ball lightning « the size of a tennis ball" has gone through a closed window, in which the glass had a crack.

Ball lightning «flowed" in the hole between the logs in the forge room. The slit width was much smaller than the ball lightning diameter. ball lightning was a ball with a diameter of 12-13 cm, orange, brightness of the lamp 50-100 watts.

Yellow ball the size of a large orange was creeping through the crack in the wall. Rather, it was not creeping but was poured from one-half to the other.

Ball lightning walked into the room through a hole in the glass, was flattened, as its size was larger than the hole. Eyewitnesses clarify: "the ball was in the 10-15 cm from our faces, and we have seen well as he began to pass through the hole, taking the form of a melon. He stretched out, was less in diameter and passed through the hole. When the ball passed through the hole and decreased in size, it was shaking all the time, and it seems that it consists of jelly".

After ball lightning has penetrated the steam line, it begins to move against the movement of steam in the direction of the fuel elements. Again, in the book of Sagan, an entire chapter is devoted to describing cases when ball lightning moves against the wind.

Once in the mine, ball lightning burned down a hole of the correct round shape in the metal wall of the mine. A description of the cases when ball lightning burns down a circle hole through a metal plate can be found in Stakhanov's book. There is evidence that ball lightning with a diameter of about 20 cm burned down a circle hole of about 7 cm diameter in the side wall of the factory pipe ([4], case 34). Besides, there are numerous evidence that the ball lightning that appears in the airplane cabin burned down holes through their skin.

Similar holes of regular round shape, burnt in targets made of electrically conductive materials (aluminum, titanium, tungsten) can be seen in the photographs shown in Fig. 22 in an article by R.F. Avramenko, V.I. Nikolaeva, L.P. Poskacheva "Energy-intensive plasma formations initiated by an erosive discharge is a laboratory analogue of ball lightning" [11]. It is noted that the burnt holes have a regular round shape without signs of fusion. Thus, the literature describes all the properties of ball lightning that are necessary to perform all 6 of the above events. Therefore, the hypothesis that the initial cause of the accident is ball lightning allows us to explain circumstances that were left without explanation and were ignored as irrelevant to the phenomena under consideration.

The probability that all these circumstances are random is close to zero. Thus, the appearance of ball lightning led to an emergency that could not be foreseen by the operators or the designers. There was a depressurization of the system, providing a constant flow of steam into the turbine. In this case, relaxation oscillations could occur.

Conclusion

You can see that there are all prerequisites to explain the simultaneous existence of the following irrefutable events

1. the start of testing, initiated appearance the following phenomena observed for 10-15 seconds

2. there was a bad sound like do-do-do-do ....

3. the repetition of this sound decreased, but the strength of the sound increased. After that there was an explosion.

4. the smell of burned transformer oil was felt

5. "a great circle of **a regular round shape**" in a sheet of the metal wall of the reactor was burnt by unknown object

6. the walls of the reactor shaft of the 4th power unit were covered with an oil emulsion".

The probability that all these 6 events appeared simultaneously by chance is an infinitesimal quantity.

There is no reasonable alternative explanation for these events.

Even if the primary cause of the tragedy was not ball lightning, it is necessary in the future to exclude the possibility of the occurrence of ball lightning at nuclear power plants. At the same time, one must not only take great care in switching high currents, but also take measures to prevent natural ball lightning from entering the nuclear power plants room just like they get into the cabin of airliners. There are hundreds of reports of such cases.

Ventilation is not permissible there an excessive pressure is created in the building of nuclear power plants. In this case, all ball lightnings appearing near the building will tend to penetrate inside just like they penetrate the cabin of a flying airplane, where the pressure is higher than outside.

References

1. Gorbachev B.I. The causes of the Chernobyl accident: the final choice between the two versions. // Collection "Problems of Chernobyl", issue 10, part 1 of Chernobyl 2002.
2. Chernobyl may recur. The mystery of a nuclear disaster is close to a solution. Russian Newspaper New Izvestia /2000/09/05.
3. Barry J. D. Ball Lightning and Bead Lightning. N.Y.: Plenum Press, 1980.
4. Stakhanov I. P. On the physical nature of ball lightning. (Moscow, Atomizdat, 1985).
5. Ed. R.F. Avramenko. Ball lightning in the laboratory. (Moscow, Chemistry, 1994).
6. Egorov A.I., Stepanov S.I., Shabanov G.D. Demonstration of ball lightning in the laboratory. Physics-Uspekhi 174 No. 1, January 2004, p. 107-109.
7. Bychkov V.L. Bychkov A.V., Timofeev I.B. Experimental modeling of long-lived luminous formations based on polymer organic materials. ZhTF 74, issue 1 (2004) p. 128-133.
8. V.P. Torchigin, A.V. Torchigin. Phenomenon of ball Lightning and its outgrowths. Physics Letters A 337 (2005) 112-120.
9. Torchigin V.P. On the nature of ball lightning. Reports of the Academy of Sciences, Volume 389, No. 1, 41-44 (2003).
10. V.P. Torchigin, A.V. Torchigin. Features of ball lightning stability. Europhysics Journal D 32 (2005) 383-389.
11. R.F. Avramenko, V.I. Nikolaev, L.P. Poskacheva "Energy-intensive plasma formations initiated by an erosive discharge is a laboratory analogue of ball lightning." In the collection Ball lightning in the laboratory. (Moscow, Chemistry, 1994).
12. A.I. Klimov, G.I. Mishin. Anomalous wave and gasdynamic properties of long-lived energy-intensive plasmoids. Letters to the PTF. Volume 19, issue 13 pp. 19-24 (1993).
13. S.E. Emelin et al. Some objects arising from the interaction of an electric discharge with metal and polymer. ZhTF 67 No. 3 19-28 (1997).
14. E.M. Pazukhin. "Explosion of a hydrogen-air mixture as a possible cause of the destruction of the central hall of the 4th block of the Chernobyl nuclear power plant during the accident on April 26, 1986", Radiochemistry, vol. 39, no. 4, 1997.
15. Torchigin V.P. On the Nature of Ball Lightning Doklady Physics, Vol. 48, No. 3, 2003, pp. 108.
16. Torchigin V.P. Optical Resonators in the Atmosphere Laser Physics, Vol. 13, No. 6, 2003, pp. 1 14.

17. Torchigin V.P. Optical Resonators in the Atmosphere Laser Physics, Vol. 13, No. 7, 2003, pp. 919
18. Torchigin V.P. Propagation of Self-Confinement Light Radiation in Inhomogeneous Air Physica Scripta. Vol. 68, no. 6, (2003) 388.
19. Torchigin V.P. Mechanism of the Appearance of Ball Lightning from Usual Lightning Doklady Physics, Vol. 49, No. 9, 2004, pp. 494.
20. Torchigin V.P., Torchigin A.V. Manifestation of Optical Quadratic Nonlinearity in Gas Mixtures Doklady Physics, Vol. 49, No. 10, 2004, pp. 553.
21. Torchigin V., Torchigin A. Role of Ball Lightnings in Low Energy Nuclear Reactions Infinite Energy 54 (2004) 46
22. Torchigin V.P., Torchigin A.V. Space solitons in gas mixtures Optics Communications 240 (2004) 449
23. Torchigin V.P., Torchigin A.V. Behavior of self-confined spherical layer of light radiation in the air atmosphere Physics Letters A 328 (2004) 189
24. Torchigin V.P., Torchigin A.V. Self-organization of intense light within erosive gas discharges Physics Letters A 361 (2007) 167
25. Torchigin V.P., Torchigin A.V. On phenomenon of light radiation from miniature balls immersed in water Physics Letters A (2010) 374
26. Torchigin V.P., Torchigin A.V. Interrelation between ball lightning and optically induced forces Phys. Scr. **88** (2013) 035402 (6pp)
27. Torchigin V.P., Torchigin A.V. Nonlinear properties of gaseous optical mediums in a context of ball lightning explanation Optik 127 (2016) 2319
28. Torchigin V.P., Torchigin A.V. Stability of the self-confined light Optik 127 (2016) 2298
29. Torchigin V.P., Torchigin A.V. How Ball Lightning penetrates in room through small holes and splits Optik 127 (2016) 6155
30. Torchigin V.P., Torchigin A.V. Ball Lightning as a self-confined light Optik 127 (2016) 2202
31. Torchigin V.P., Torchigin A.V. Nonlinear properties of gaseous optical mediums in a context of ball lightning explanation Optik 127 (2016) 2319
32. Torchigin V.P., Torchigin A.V. How the ball lightning enters the room through the windowpanes Optik 127 (2016) 5876
33. Torchigin V.P., Torchigin A.V. How ball lightning manages to catch up a flying aircraft and penetrate into its salon Optik 148 (2017) 196
34. Torchigin V.P. Explanation of abnormal behavior of ball lightning near the earth Surface Optik 171 (2018) 167
35. Torchigin V.P. Explanation of anomalous bouncing luminous droplets of liquid silicon in a framework of the optical model of Ball Lightning Optik 171 (2018) 188

36. Torchigin V.P. Explanation of abnormal motion of ball lightning near the earth's surface Optics 171 (2018) 149

37. Torchigin V.P. Explanation of abnormal motion of glowing silicon balls in a framework of optical model of ball lightning Optik 176 (2019) 704

38. Torchigin V.P. How optical model of ball lightning explains its paradoxical movement upwind Optik 184 (2019) 533-537

39. Torchigin V.P. Why ball lightning disappears suddenly traceless Optik 187 (2019) 65-69

40. Torchigin V.P. How ball lightning finds out slots and holes to penetrate through them Optik (2019) 163126

41. Torchigin V.P. Ball lightning as a bubble of self-confined light Optik 186 (2019) 63-71

42. Torchigin V.P. Physics of a ball lightning in a form of a bubble of light Optik 188 (2019) 294-301

43. Torchigin V.P. Ball lightning as a bubble of light: Existence and stability Optik 193 (2019) 162961

44. Torchigin V.P., Torchigin A.V. Simple explanation of physical nature of ball lightning Optik 203 (2020) 164013

APPENDIX

We present additional information about the properties of ball lightning, obtained not from eyewitness accounts, but on the basis of well-known laws of physics under the assumption that ball lightning is circulating light. This information allows us to more clearly understand such properties of ball lightning as the absorption of transformer oil vapor, the beneficial effect of evaporation of transformer oil on the occurrence of ball lightning. Of particular note is the simple and natural justification for the possibility of movement of ball lightning against the movement of steam. We draw attention to the fact that all the above anomalies and properties of ball lightning are confirmed in the optical theory of ball lightning. This allows us to relate to the conclusions of this theory with great confidence.

What is ball lightning

Ball Lightning is a symbiosis of a thin spherical layer of strongly compressed air and an intensive light that circulates in the layer in all possible directions. They help each other.

The thin layer of the strongly compressed air is the lightguide that prevents radiation of the light in surrounding space. In turn, the intense circulating light compresses the air due to the phenomenon of the

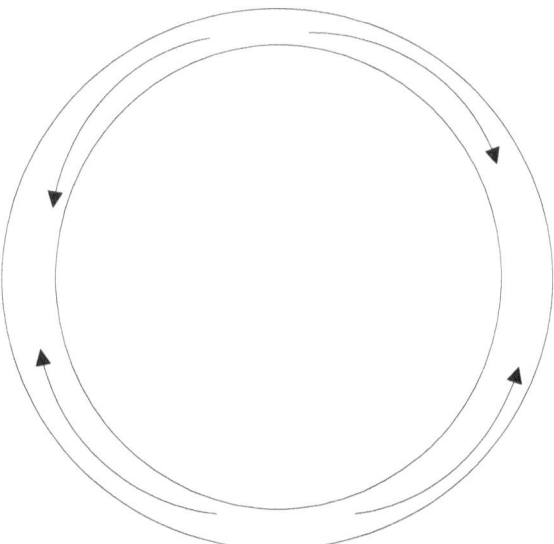

Fig.3. Cross section of a thin spherical layer of compressed air by a plane passing through the center of the sphere. The circulating light is rotating in the layer in all possible directions.

electrostriction pressure. White light fells into a trap that it created for himself.

A cross-section of Ball Lightning by any plane that passes through the center of the sphere is shown in Fig.3. The light is circulating in a thin film of strongly compressed air. That is all.

This short explanation can be clarified a bit. The air pressure inside and outside the spherical layer of compressed air is the same and equal to atmospheric pressure. Unlike a soap bubble, in which its spherical shape is provided by excess pressure inside the soap film, the spherical shape of the light bubble is provided by centrifugal pressure on the film produced by the light circulating in it.

Since the light bubble glows, that is, it continuously emits light into the surrounding space, the amount of light stored in it gradually decreases. When this amount becomes less than a certain threshold, the light bubble becomes unstable and bursts.

The lifetime of a light bubble is approximately 4 orders of magnitude longer than the lifetime of white sunlight in the earth's atmosphere. The first publication in 2003 showed that this situation can only happen if the density of compressed air is extremely high and approaches the density of liquid air.

The intensity of the light circulating in the shell increases by approximately one billion times in comparison with that of the same light that propagates in a straight line. This is due to the fact that the light makes approximately one milliard revolutions per second, that is, it crosses the same cross section one billion times per second. As a result, the energy density in the shell also increases by one billion times. This explains the abnormally high energy of ball lightning and the extremely high density of compressed air in the shell.

Of course, in the shell of a bubble of light can be not only air, but also other gases. It is shown that the shell draws gases from the atmosphere surrounding it, the refractive index of which is greater than the refractive index of air

Thus, circulating light or a bubble of light or a ball of light is a completely new object that has not been studied previously either theoretically or experimentally.

The unique physical conditions in the shell of a light bubble can find application in the creation of new alternative energy sources, but this is another topic.

Properties of the circulating light

Let us first consider well known properties of ordinary light propagating in a straight line. It is correct for vacuum and a homogeneous optical medium. In an inhomogeneous optical medium, a beam of light is curved in the direction where the reflective index of the medium is increased. The most famous example of curvilinear propagation of light is mirages in deserts, when an observer can see objects beyond the horizon. This can take place only if the light from the object to the observer's eyes propagates along a curve of the trajectory, as shown in Fig. 4.

Fig.4. The top of the building is seen when a light bean propagates in the homogeneous air. The whole building is visible when the beam propagates in the air of inhomogeneous density. In this case the light bean deviates from a straight line at a distance equal to the height of the

Let us estimate the beam shift in the transverse direction under the assumption that the observer sees the object at a distance of 30 km. As you can see from Wikipedia, the horizon for the observer is 4.7 km. At a distance of 30 km, the observer with a straight-line propagation of light sees the top of an object with a height of 50 m. If in the desert the observer sees the whole object, then the beam at a distance of 30 km is deflected by 50 m. Since at a speed of light of 300,000 km / s the light travels a distance of 30 km behind 0.1 ms, then the beam velocity in the transverse direction is 500,000 m/s = 500 km/s.

Let us now consider the idealized case when a beam of light turned from a vertical reflector in Fig.3a located at a distance equal to half the distance

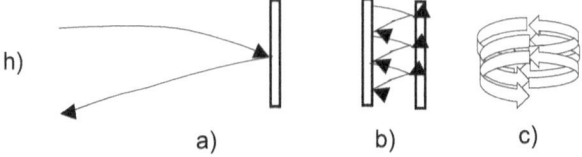

h)

a)   b)   c)

Fig. 5. Comparison of deflection of the light beam in an inhomogeneous air for a) one vertical reflector? b) two vertical parallel reflectors, c) circulating light without reflectors

from the observer to the building in Fig. 2. In this case deflections h in Fig.2 and Fig.5a are identical on assumption that inhomogeneities of the air in the first and second parts of the path in fig.4 are identical. The same deflection we will have if the light beam reflects periodically in the same inhomogeneous air from two ideal reflectors in the form of two parallel vertical planes as is shown in Fig. 5b. If the optical medium between the planes is identical to that in the desert, the light beam will need to move downward at a speed of 500 km/s.

Consider now the circulating light shown in Fig.5c instead of the light between the reflectors. In this case, reflectors are not required, since during rotation the light itself periodically changes its direction of propagation. In this case, the light will also shift downward. We can say that the circulating light is screwed into the optical medium in the direction of increasing the refractive index. For example, if the diameter of the circle around which the light rotates is 10 cm, then the light makes about one billion revolutions per second. Since at the same time it shifts by 500 km, then it shifts by 0.5 mm per one revolution.

Of course, the heterogeneity of the atmosphere under ordinary conditions is much less than the heterogeneity of the atmosphere in the desert and the speed of movement of circulating light under ordinary conditions is correspondingly lower. For example, if the speed of movement of the circulating light is 1 m/s, then the inhomogeneity of the atmosphere should be reduced by 500,000 times at this speed. We can conclude that circulating light is an extremely sensitive instrument for detecting the slightest inhomogeneity of the atmosphere.

The circulating light is located in the maximum of the air density and is moving in space with this maximum.

To say that the circulating light is extremely sensitive to changes in air density - this means to say nothing. The sensitivity of the circulating light is such that it senses a change in air density near the window of the building, in which the temperature is lower than outside. The thermal conductivity of the window is higher than the thermal conductivity of the walls of the building, so the air is colder on the outside of the window than on the outside of the wall. The density of air increases with decreasing temperature. This increase feels the circulating light and goes to the window, and not to the walls of the building.

However, if there is a small hole in the wall of the building between the room and the outside, then the air outside this hole is colder than that near a solid wall. In this case, the air density near the hole is greater than near the wall. In this case and the circulating light is directed to the hole.

Heterogeneity of the air atmosphere

Since the circulating light feels the slightest changes in the refractive index of the air, where it is located, we consider the reasons causing such changes. Since the refractive index of air linearly depends on its density, for a qualitative analysis it is enough to analyze the distribution of air density in space. In turn, the density of air of uniform composition (without impurities of other gases whose concentration varies in space) is directly proportional to air pressure and inversely proportional to air temperature, expressed in degrees Kelvin. This information is enough to analyze the slightest changes in the density of air near the surface of the earth.

1. Distribution of the air density along the height above ground

It does not require additional explanation that the air density decreases with increasing height. Moving in the direction of increasing air density, the circulating light moves toward the surface of the earth. But it does not touch the surface, since the indicated property of the atmosphere is violated near the surface itself. This is due to the fact that directly near the surface of the earth is warm air, which is heated from the surface of the earth due to the phenomenon of heat conduction. In turn, the surface of the earth is heated by solar radiation.

It is known that the density of air decreases with heating. For example, the air inside a balloon is heated by a burner. Its density decreases and the weight of the air inside the shell of the balloon decreases. The weight of this air, together with the shell of the balloon, becomes less than the weight of the cold air displaced by the balloon. According to the Archimedes law, the ball rises up to the height where these weights become equal.

Thus, the air density gradually increases at approaching the earth. However, directly near the surface of the earth, the air density gradually decreases. This means that somewhere near the surface of the earth there is a maximum of the air density. At this maximum, the circulating light, which moves in the direction of increasing air density, cannot move either up or down. The circulating light stops at the height where the air density is maximum. According to observations of motion of natural ball lightnings, this height ranges from 0.5 to 5 meters.

Incidentally, the following fact indicates that the air directly near the earth is warmer than at a higher altitude. Anti-fog headlights on cars are placed as close to the surface of the earth as possible. This is because the air temperature is higher immediately near the surface of the earth, and fog does not form at high temperatures. Thus, the anti-fog headlights do not disperse the fog, but shine in the place where the fog is absent since the temperature in this place is greater.

2   Distribution of the air density in horizontal direction along the surface of the ground

Since the circulating light is a sensitive device that responds to the slightest changes in the air density and moves in the direction in which the air density increases, it is necessary to analyze what can affect the change in the air density. First of all, these are buildings. People try to live at a temperature that is comfortable for them, and not at the temperature that nature provides them. This explains the widespread use of various heaters and air conditioners. As a result, the temperature of the walls of buildings differs from the ambient temperature. This is especially noticeable near various heterogeneities the walls such as holes, windows, open windows.

In addition, it must be borne in mind that the heat capacity and thermal conductivity of the surface of the earth, air, and buildings are different. As a result, they react differently to the 24 hour cycle associated with the rotation of the earth around its axis.

For example, buildings have a foundation buried several meters in the ground. The temperature at this depth is close to the average annual temperature and varies slightly during the day compared to changes in the temperature of the roof and walls. At the same time, due to the phenomenon of thermal conductivity, all components of the building are interconnected.

In addition, the circulating light itself affects the temperature of the air. This air is transparent and cannot be heated by light that radiates the circulating light due to the phenomenon of molecular light scattering. The radiated light is visible to the observer of natural ball lightning. However, the situation changes if the circulating light is near some obstacle. The obstacle absorbs the radiated light and its surface heats up. The heated surface heats nearby air layers and the air density in these layers decreases. The circulating light must respond to this circumstance.

This simple and indisputable information is enough to explain all the anomalies, secrets and intriguing behavior of natural ball lightning.

Explanation of ball lightning riddles on assumption that it is the circulating light

1 Why the erosive gas discharge is favorable for arising ball lightning

According to the optical model, the circulating light circulates in a thin layer of highly compressed air. The refractive index of air when it is compressed increases. The layer with an increased refractive index is a light guide, which prevents the radiation of light into free space. In the same time an increase in the refractive index can be achieved in another way by means of absorption of molecules of other gases from the surrounding thin layer of space, the refractive index of which is greater than the refractive index of air.

This method of producing ball lightning was first discovered by experimenters in an attempt to obtain ball lightning in laboratory conditions. It was noted that anomalous objects are obtained more often if the electrodes are coated with a layer of hydrocarbons. It was noticed In the spectral analysis of the obtained objects that the anomalous objects contain molecules of these hydrocarbons.

It has been shown theoretically that the absorption of molecules with an increased refractive index requires less energy to increase the refractive index than simple compression of air. Therefore, the resulting objects draw molecules of surrounding hydrocarbons into their shell. In addition, the refractive index of a gas mixture of air and hydrocarbons is greater than the refractive index of air. Since the emerging object tends to move in the direction of increasing the refractive index, it remains in the discharge region during the erosive gas discharge. In this case, the accumulation of light in the shell of the object takes a longer time. This contributes to obtaining objects with a longer lifetime.

2 How ball lightning can burn out a hole through a metal plate.

When ball lightning approaches an obstacle, it heats it up due to its radiation of light. The obstacle heats nearby air layers due to the phenomenon of thermal conductivity. The refractive index of warm air decreases and ball lightning bypasses the obstacle.

The situation changes if the obstacle has a high heat capacity and thermal conductivity, for example, is a metal. In this case, an attempt to heat the metal fails. The heat received by radiation propagates through the metal and the heating of the metal is minimal. Therefore, ball lightning can get close to metal.

The intense circulating light begins to evaporate the metal. Then erosion of the metal occurs and molecules of the metal are absorbed in the ball lightning shell. The refractive index in the region between the ball lightning and the metal increases. Located in this area, ball lightning evaporates the metal until it burns a hole through the plate, or until its energy runs out. In the latter case, a crater is observed in the metal.

3 How can ball lightning change its spherical shape to penetrate into any hole smaller than its diameter

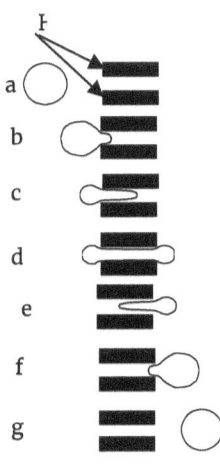

Figure 6. Steps of penetrating BL through a hole in a wall

Approaching a hole whose transverse dimensions are less than the diameter of circulating light, it begins to heat the walls adjacent to its surface. The air temperature near the walls increases, and the air density decreases. Circulating light is in the area where different parts must move in different directions. This leads to deformation of the circulating light in such a way that the parts of the circulating light adjacent to the walls repel from the walls. This effect leads to the fact that the surface of circulating light cannot touch the wall. The areas of the circulating light that are remote from the walls continue to move toward increasing the air density and pull after the whole circulating light as is shown in Fig.6.

4 Why can ball lightning move against the wind?

Scientists are even more perplexed by the ability of ball lightning to move against the wind. There must be some kind of engine that allows ball lightning to overcome wind resistance. Only creatures and airplanes have this ability.

When the circulating light moves against the wind, air flows around it and tries to slow down. As a result, the air pressure on the front hemisphere is higher than on the back. Then the air density on the front hemisphere is greater than on the back. In accordance with the main property of the circulating light to move in the direction of increasing air density, the circulating light moves against the direction of the wind.

In our case, there is another reason for the movement of ball lightning against the flow of water steam. Since water steam moves in a limited space through the steam line from the separator 6 to the turbines 8, the steam pressure near the separator is greater than that near the turbines. In this case, the steam density, and therefore the refractive index near the separator, is greater than that near the turbines. Since the main property of ball lightning is to move towards an increase in the refractive index, ball lightning will move in the line steam 7 towards the separator, and from there towards the reactor in the direction to the reactor 3 where the temperature decreases.

5 Why ball lightning doesn't hit the ground when falling from the sky

Many scientists are perplexed by the following fact. All ball lightnings that fall to the ground from heaven do not hit the ground, but slow down and stop at a certain distance from the surface of the earth, as if they would be afraid of breaking on the ground. This property of ball lightning distinguishes them from all other known objects. Whatever internal processes occur inside ball lightning, its behavior is determined by external forces. These forces should somehow distinguish ball lightning from all other known objects that, when falling from the sky, hit the ground. These forces should begin to slow down ball lightning so that its speed decreases to zero at a distance of several meters from the ground. Several questions immediately arise. What is the nature of these forces? Why do they begin to act only directly near the surface of the earth? What is the difference between this distance and other distances?

It can be assumed that ball lightning has a jet engine, which is turned on to brake directly around the earth. In addition to the question of how such an engine is designed, other questions arise. How does this engine work? What instrument, like an altimeter, do ball lightnings use to determine the distance to the earth? What tools do ball lightning include to slow down to zero and not hit the ground? It would seem impossible to find reasonable answers to these questions; however, all questions receive their answers if we admit that ball lightning is a circulating lightro

The answer is much shorter than the questions. The circulating light moving in the direction of increasing air density stops at the height where the air density is maximum. It cannot move either up or down, as the air density decreases in these directions. We previously showed that the maximum of the air density is located at a distance of several meters from the earth's surface.

Why does ball lightning move horizontally at some distance from the ground?

Why doesn't it fall to the ground if it is heavier than air? Why doesn't it rise if it is lighter than air? What forces make it stay near the surface of the earth?

The answer is simple. The maximum air density in the vertical direction is at some distance from the earth's surface. The circulating light moves at this height.

6 How does ball lightning feel obstacles?

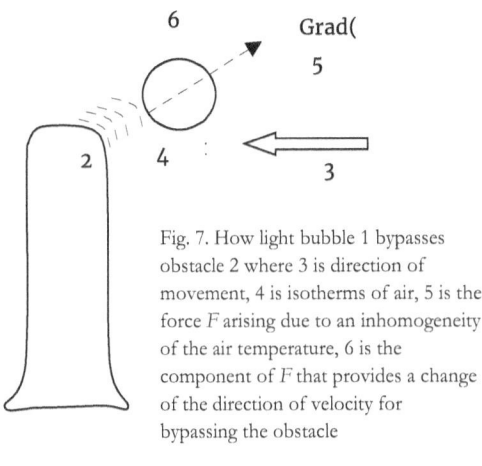

Fig. 7. How light bubble 1 bypasses obstacle 2 where 3 is direction of movement, 4 is isotherms of air, 5 is the force F arising due to an inhomogeneity of the air temperature, 6 is the component of F that provides a change of the direction of velocity for bypassing the obstacle

How does ball lightning feel that an obstacle has appeared in its path and what means it uses so that it does not run into the obstacle, but bypasses it?

Approaching an obstacle, the circulating light heats it due to its radiation. The obstacle warms up nearby air layers. As a result, air density decreases when approaching an obstacle. Once in this zone, the circulating light moves in the direction of increasing the air density, that is, from the obstacle as is shown in Fig.7.

7 How does ball lightning detect holes to penetrate through them into the room?

It is known that ball lightnings penetrate various openings and crevices into the rooms. Several questions immediately arise. How ball lightning finds these openings. What sensory organs are used. Further, what means does ball lightning use to organize its movement towards the opening?

In the event that ball lightning is circulating light, the answer to this question is simple.

Getting into the area near the hole, the circulating light enters the inhomogeneity of air density. Heterogeneity is created by the hole. If the air temperature in the room is lower than outside, the density of the air outside the room gradually increases as you approach the hole. Moving in the direction of increasing air density, the circulating light moves toward the hole. The density of the air in the hole itself increases as you approach the room. Moving in the direction of increasing air density, the circulating light enters the room through the hole.

8   How can ball lightning penetrate glass panes?

Scientists scratch their heads in perplexity. How can ball lightning penetrate glass panes without damaging it? What then does it consist of? We are well aware that no material particles can penetrate through glass. What then emits light?

The window is transparent to light, and light can penetrate through the glass. Windows exist for that, so that light penetrates through them. The circulating light freely penetrates through the glass transparent to it. Compressed air cannot penetrate the glass, but this is not required. Exactly the same air is on the other side of the glass, which is compressed by the penetrating circulating light.

Stages of penetration of the Ball Light shell through the window pane are shown in Fig.8.

Letter P in Fig.8 shows the region where a compression of air on the outer side of the glass takes place. The air density increases and a direction of propagation of light in the region is changed in such a way that the light returns backward in the shell through the glass.

Direction of increasing of the air density

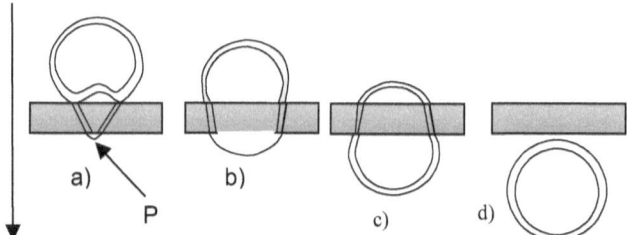

Fig.8. Stages of penetration of the Ball Light shell through the window pane
a) initial stage of penetration when the region of the compressed air is produced on the back side of the window pane
b) intermediate stage when a smaller part of the Ball Light shell penetrated on the back side in the room
c) intermediate stage when a greater part of the Ball Light shell penetrated on the back side in the room
d) Ball Light penetrated in the room completely

9 How can ball lightning accompany flying airliners and penetrate into their salons?

First, how do it discover a plane. Military experts would have given a lot to reveal the secret.

Secondly, how can it catch up and disarm the plane? What engines and fuel do it use?

Thirdly, why ball lightning is not blown away by a headwind, whose speed exceeds the wind speed in a severe hurricane? Questions may be continued.

An answer is simple. A flying airliner creates a large heterogeneity in the air density around itself. Air pressure near the wing of the airliner increases by 20%. The air density increases accordingly. This heterogeneity moves with the airliner. The air density increases with approaching the airliner. The circulating light that has got into this heterogeneity moves in the direction of increasing the air density, that is, in the direction of the airliner and catches up with it

10 What are the sources of white light that ball lightning emits throughout its existence if there are no particles in it

Indeed, there are no excited atoms that emit photons in the circulating light. The nature of the white light emitted by the circulating light is different from the nature of the white light emitted by a body heated to a high temperature. The white light emitted by the circulating light is its part. Light is emitted due to the phenomenon of molecular light scattering. This is the phenomenon responsible for the blue color of the sky and the white color of the clouds.

11 Why does ball lightning disappear suddenly at an arbitrary stage of its movement, leaving no trace after its disappearance?

What then did it consist of. Perhaps it just becomes invisible or is hiding somewhere?

Two words is enough to explain this phenomenon. This is the "occurrence of instability". Like a soap bubble that bursts when the film thickness becomes less than a certain threshold, the circulating light disappears when the light intensity, which gradually decreases due to light emission, becomes less than a certain threshold. Light scatters in all directions. Compressed air expands and there are no traces left in the place where the circulating light was located.

Natural ball lightning are the circulating light

Everyone had the opportunity to make sure that the behavior of the circulating light based on a simple law, according to which the circulating light moves in the direction of increasing air density, and the behavior of natural ball lightning, based on numerous eyewitness accounts, are completely identical. Everyone must decide whether the circulating light exist in nature. We only note that the knowledge available in the 19th century was quite enough to discover the nature of ball lightning.

Believing a priori, that the circulating light or bubble of light is an objective reality, we have shown that in accordance with well-known laws of physics behavior of bubbles of light in the conventional air atmosphere is completely identical to mysterious and paradoxical behavior of natural ball lightning. Too many puzzles related to ball lightning can be explained by the assumption that ball lightning is a bubble of light. We have shown that our bubble of light has simultaneously all listed properties. It is unlikely that this is a mere coincidence.

In addition, we are trying to explain shortly in a popular form a relationship between the circulating light and other known optical devices. The circulating light can be considered as a generalization of optical incoherent spatial solitons, in which the curvature in two mutually perpendicular directions is nonzero.

From the presented consideration it is necessary to draw a conclusion that bubbles of light are not game of mind, but it is an objective reality. Till now bubbles of light were studied neither theoretically, nor experimentally. Anybody did not suspect at all about an opportunity of their existence. Ought to underline that the existence of bubbles of light in nature is based not on theoretical rationale but on wonderful coincidence of behaviors of bubbles of light and ball lightning in air atmosphere as well as on their common physical properties.

At the same time, a theoretical analysis of the optical model of ball lightning made it possible to learn about the value of air pressure and the density of light energy in the shell of ball lightning. It turned out that the air pressure is comparable to or exceeds the maximum pressure of the gases currently obtained in laboratory conditions. The light energy density is several orders of magnitude higher than the energy density in known energy storage devices.

Conclusion

The above properties of natural ball lightning, which contributed to the occurrence of ball lightning and the disruption of the normal operation of a nuclear power plant, are confirmed by consideration of the theory that the nature of ball lightning is optical. This theory explains the mysteries and abnormal properties of natural ball lightning, including those that we mentioned when considering a possible accident scenario. This is the only theory that explains all the anomalies in the behavior of natural ball lightning. This circumstance allows us to state that this theory correctly describes the nature of ball lightning.

.

www.ingramcontent.com/pod-product-compliance
Lightning Source LLC
Chambersburg PA
CBHW030546220526
45463CB00007B/3005